THE COUNTRY TREASURY

———

THE SHEPHERD

THE SHEPHERD

JOHN SEYMOUR

Illustrated by Sally Seymour

SIDGWICK & JACKSON
LONDON

First published in Great Britain 1983
by Sidgwick and Jackson Limited,
1 Tavistock Chambers, Bloomsbury Way,
London WC1A 2SG.

Conceived and produced by
Shuckburgh Reynolds Limited,
8 Northumberland Place, London W2 5BS.

Designed by Nicholas Thirkell & Partners

Typesetting by SX Composing Limited
Printed and bound in Spain by
Printer Industria Grafica, Barcelona
DLB40830ᐟ1983

ISBN 0–283–98922–X

—I—

MOUNTAIN SHEPHERDS: BREEDERS OF EWES AND RAMS

Page 17

—II—

LOWLAND SHEPHERDS: MIDWIVES AND NURSEMAIDS

Page 67

—III—

MAN AND DOG: A RITUAL OF GRACE AND BEAUTY

Page 97

We shepherds are the best of men that 'ere trod English ground.
When we come to an ale-house, we value not a crown,
We spend our money freely, we pay before we go.
  There's no ale
  On the wolds
  Where the stormy winds do blow!

That man would be a shepherd must have a valiant heart.
He must not be faint-hearted, but boldly play his part.
He must not be faint-hearted, be it rain or frost or snow.
  There's no ale
  On the wolds
  Where the stormy winds do blow!

When I kept sheep on lofty hills, that made my heart to ache
To see the yows hang out their tongues, and hear the
  lambs to blate,
So I plucked up my courage, and o'er the hills did go,
  Penned them
  In the folds
  Where the stormy winds do blow!

As soon as I had enfolded them I turned my back in haste
And went straight to the ale-house, good liquor for to taste,
For drinks and jovial company – they are my heart's delight
  While my sheep
  Lie asleep
  All the fore part of the night!

Gloucestershire Shepherds' Song

INTRODUCTION

A N ANIMAL that has accompanied mankind on his journey from paleolithic times right down to the present, and has been with him in the hottest and driest deserts in the world, on high mountains, in Arctic regions, in the wettest marshes, and on the best of land, both arable and pasture, can surely be called versatile. An animal that has been domesticated so long – probably longer than any other – must have lost most of its wild characteristics along the way, and this we must suppose the sheep has done.

Almost wherever I have lived, if I could only have had one domestic animal, I would have chosen sheep. The sheep provides us with incomparably the best clothing material in the world. He provides us with good meat. He provides us with the best of milk. Roquefort, that most excellent of cheeses, is made of course from sheep's milk as are so many high class French, Swiss, Italian and Greek cheeses. The keeping of Friesland milk sheep is becoming a craze in southern Germany, Austria and Switzerland, and is even beginning to take root in England, that land of many sheep breeds. (The British Isles, incidentally, contain more distinctive breeds of sheep than the rest of the world put together.) I have drawn nearly half a gallon of fine milk from a Friesland sheep in Switzerland, high up in the Alps, at the home of a friend of mine.

My own experience of sheep has been nothing if not diverse. As a boy, I spent a month right out in the middle of the Romney Marshes in Kent, at a place called Scotney Farm, working as learner-assistant to a "looker", as shepherds in that part of the country were then called.

It was during the lambing season, and I assisted in helping six hundred ewes to lamb. The Romney Marsh sheep is a fine breed, which has spread all over the world, even to Russia and China, where it is used to improve many varieties of native sheep. The Romney Marsh itself in those days was all under grass – some called it the finest pasture in the world – and sheep were the only stock kept in any numbers. Now, alas, much of the Marsh has been ploughed up, and that great flat expanse, reclaimed from the sea by the Romans, is criss-crossed with electric pylons.

But when I worked there it was sheep and nothing but sheep, and sheep as far as you could see. I slept in the looker's house – where I was fed magnificently by his wife – when I did sleep, but much of the time I did not sleep at all, because the looker and I would sit most of the night around a hot coal fire in his "ship-hus" or sheep house: a tiny brick hut right out on the marshes, generally with two or three wet little lambs warming up by the fire and being resuscitated with warm milk from a bottle. Every hour or so we made sorties out into the cold wet windy dark: to help ewes in their labour, to make sure the little new-born babies had suck, perhaps to skin a dead lamb and fit the skin over one of a live pair of twins so as to fool the mother of the dead lamb into adopting it.

I learned what it was to kneel on the sodden ground in the dark, with the rain sluicing down my neck, and force my arm – well lubricated with carbolised oil – so gently into a ewe's resisting uterus to feel what was happening in there – why a pair of twins, perhaps, was refusing to be born; and then, working in that inexorable pressure of the womb, to sort out the little limbs, gently draw out first one lamb and then the next, generally alive and soon to be jumping and kicking; and then tenderly and carefully to ensure that the mother got up and allowed the weak little wet lambs to suck. The sight of a ewe forgetting the pain of birth to turn round and gently lick her tiny lamb is one of the most beautiful and tender things in nature.

My next encounter with sheep was on the Cots-wold Hills. I worked there as a sheep-minder for Arthur Garne Esq, of Cocklebarrow Farm, Alds-worth, for six months of one winter. My job was to fold a couple of hundred wethers (castrated male sheep) and about sixty ewes of the Cotswold breed, on turnips. This was quite a different experience, and involved moving forward heavy hurdles – made from riven ash – every day so as to construct new folds, or pens, for the sheep to run into the next day. This was heavy labour, and was not made easier when the ground became frozen hard and I could not get the "sheep-bar" as we called it (it's called a "fold-pritch" by Suffolk shepherds) into the ground so that I could drive in the stakes, or shores, to hold the hurdles upright. Besides moving the hurdles I had to take hay, corn and water to the sheep, and trim any feet that had footrot, and generally look after these delicate animals. I think I developed that year a little of the love-hate that shepherds everywhere seem to bear their charges.

I worked with sheep again in the Karoo, in South Africa. There it was my job for six months to "ride camp" on the farm of J. Oscar Southey, who was

at that time probably the premier producer of high quality Merino breeding stock in Africa.

Southey had about thirty thousand sheep on some uncounted number of acres. The land was divided up into paddocks called "camps", each from five to ten miles across. Day after day I had to ride a horse over these camps, with a pair of shears and a bottle of disinfectant in my saddle-bag, and round up the sheep in the camp so that I could count them. If I saw a fly-struck sheep I would ride my horse into the mob and then leap off and grab him, scrape the seething maggots off with the shears, shear off the dirty wool, and douse on the disinfectant. If left alone our woolly sheep will be eaten alive by maggots, which shows how far they have come from the tough wild mountain animals they originally were.

The Karoo is marvellous sheep country – vast plains covered with little bushes a foot or two high, bearing foliage delightful to grazing animals. There are no trees or grass: just little shrubs; and the hard even ground was delightful to canter over. The Karoo then rivalled Australia as the supplier of fine wool to the rest of the world.

The next sheep I looked after had no real wool on them at all. In that area of Asia to the east of Persia and the north of Afghanistan there is a breed of desert sheep called Karakul; these are kept for the skins of their new-born lambs more than for any other reason. The settlers in the former German colony of South West Africa, now generally called Namibia, imported some of the hairy black karakul rams and used them to up-grade the local native sheep, which had very short white hair like a goat, and black heads. Both karakuls and native sheep were typical desert animals. That is to say they had wide fat tails which, by the end of the rainy season when the sheep had been feeding well, contained several pounds of fat, fat which was used up before the next rains began by the sheep in its efforts to keep alive. After about three crossings with a black karakul ram the lambs were mostly born black.

The native sheep of South West Africa were owned mostly by members of that mysterious race the Hottentots, or else by the Ovambos in the north of the territory. I managed a farm with two thousand sheep on it right on the edge of the Namib Desert, twenty miles away from the nearest white man. This was quite a different sort of shepherding: in fact far more akin to the archetypal shepherding we read about in the Bible.

Stone-age men who had maintained their way of life into the twentieth century were our shepherds. They brought their flocks back from the unfenced bush-veld every night to drink water and be confined in *kraals*, pens made of wire or thorn bush. If the sheep had been left out at night, leopards and other big cats would have got them. In the morning each shepherd would let his sheep out (while I counted them) and then follow them miles out into the bush to find what fodder they could. The lambs which had been born in the bush the day before – and those dropped during the night – I would examine and decide what to do with. Good black ewe lambs would be kept as replacements for the flock. Ram lambs with white patches on them, being no good for their skins, would be spared to be killed later for their meat. But all-black ram lambs would have their throats cut, there and then, and their skin removed to adorn, ultimately, the backs of wealthy women in Paris, London and New York. If you left the lambs more than a day or two before killing them the tiny curls, which were the desired embellishments, would open out and the pelts would lose value. The lamb slaughter, I must admit, sickened me, but it interested me enormously to see how the versatile sheep had become adapted to yet another environment: the desert.

The other day I watched some friends riding their ponies and working their marvellous dogs to round up two thousand little Welsh Mountain sheep a thousand feet up the Cambrian Highlands, to dip them and draught out ewe lambs to go to the low country to spend their first winter. The sheep industry in the British Isles is a complex mechanism, driven by trade between one kind of sheep farmer and another. Few, if any, serious sheepmen – those that earn their living primarily from sheep – manage to exist without trading, directly or indirectly, with other sheep farmers: hardly anyone breeds all his own rams and ewes and fattens all his own lambs for sale.

Most ewes begin life on the mountain and hill farms of Scotland, Wales and the north of England. They spend three or four years there and are then sold at a number of celebrated annual sheep auctions to lowland farmers, most of whom come from the south of England. Rams, too, are mostly bred in the northern hill country, and then make the great trek south.

The migration of sheep from the rugged hills of the north to the lush pastures of the south is a major strand in the story this book has to tell. Suffice it to say here that the shepherds of the upland regions of Scotland, Wales and the north of England live a very different life and work towards essentially different ends from those of the lowland shepherds of the south. I have therefore given a chapter each to the two types of shepherd, and my third, and final, chapter celebrates that wondrous, intelligent, indispensable companion, common to all shep-herds: the sheep dog.

I

MOUNTAIN SHEPHERDS: BREEDERS OF EWES AND RAMS

LET US BEGIN our story in the high country – in the rugged Pennine Hills in North York-shire, where sheep are kept thousands of feet above sea level – for it is there that so many sheep start their lives, before they begin their trek down to the kinder lands and climate of lower altitudes.

"You have to be pretty sure about it, so you listen to the forecast and check. You've got to watch the barometer an awful lot and look at the sky. You've got to be very careful about the weather because you sit down here and you think – it looks a bit wintry up there – but it's a *completely* different world! You can go up there and there might be sheep under snow. We've had quite a lot under snow – you know, under the drifts."

Thus spoke Dick Fawcett – son of Richard Fawcett – as we sat in front of his house in the green valley of Wensleydale, not far from the little town of Hawes. At Hawes Dick's father runs a sheep market which is known throughout the country for the sale of the famous Mule Sheep. Dick waved his

hand towards the blue mountains in the distance –
the fell he was talking about was eight miles away
and, indeed, looked like another world.

"I have one dog which scents them, which is a
great performance. She has saved an awful lot of
lives. She digs in and then we dig 'em out. If there's
a chance of a storm, you get up there and fetch them
down to the barn in the pasture where we can be
sure of it. If you don't get 'em off in time you can get
in a real old mess. Last year for instance we had an
awful lot of snow and I went up on the Saturday
morning and we had them at the bottom of the fell,
and I thought, 'well I'd better put them further

down'; which I did. Just to be on the safe side.

"And we planned to go and stay with some friends at Welwyn Garden City for a few days, so I said to Anne, 'Let's be going: if we don't go now we never will be going.' So off we went. We went down there for about two days and I was sent an SOS saying come home! There are about a dozen sheep under the snow, and they had taken my bitch with them, and she found six or seven, but she kept looking around as if she wasn't happy with them. So I had to come home! I took her round and we found them all. These were sheep of some value. Some of them sixty or seventy pounds you know. They were worth saving."

Dick Fawcett is a very modest young man, and will not make much of the business of going up into the high mountains and digging sheep out of the snow. But I have spent some time going round the fells and dales with a neighbour of his, Willy Calvert, of the next valley – Swaledale, near Muker, the very home and heart of the cult of the Swaledale breed.

Mr Calvert is a tremendous enthusiast for the Swaledale sheep, but on one occasion he was comparing its performance in blizzard conditions with that of the Scots Blackface:

"A Scotty ewe will stop out on mountain. If

there's no one with her and it snows, she'll stop out on mountain. She finds a little place she thinks is safe – she might be right and she might be wrong. But to her this is home. This is safe. And she'll huddle in there to escape storm until it has gone, and then she will creep out and just scratch around where she is – you know just scratch around where she is. Swaledale won't do that – she'll have to be down with the others – down down down.

"The Scottish shepherds are not gathering the sheep. They can have a blizzard and never go near them see – they rely on them making their own little places. They have a lot of big rocks and old Scotty can get behind a big rock – half the size of that big hut down there – and she can find shelter; she get semi-buried but she can get out. She survives. Swaledale wouldn't be happy there – she wants to get with a bunch and might end up with a worse place and they are not as safe left alone as those Scotties were. That's probably the disadvantage, but then there again they ought to want to start farming them and gathering them up and know where they have them shouldn't they? But they have vast areas and they are not used to doing that you see. That's the answer to that! It's like sheep are out on the moor and the shepherds see it's snowing – they sit there smoking their pipe like 'aye it's snowing again'. Tomorrow there's a blizzard, 'Oh them will be all right! They're all right!' Well, yes *some* are all right – *some* aren't. They expect their losses. He maybe doesn't see them until it comes round to lambing time. 'They are due to lamb – I will go and have a look round!' He maybe see them sheep once a week in lambing – if they don't lamb and get going they are wrong aren't they? They are *used* to loss in lambs – they are *used* to loss in sheep. Well it has always been that way."

I remember when Willy Calvert was telling me this we were standing high up on a windy fell not far from The Highest Inn in England. He pointed down the sweeping hillside to a sharp little valley with a stream in it. Down there were circular enclosures of limestone. It was into these enclosures that the Swaledale sheep would creep, be herded, at times of blizzard and snow. That is, if they were lucky – or if their shepherds could find them.

Every year there is snow on the high mountains of the north of England, Scotland and Wales. Some years there is yards of it. But it is the wind which accompanies the snow which is the difficulty. The high areas of the British Isles are the climatic extreme at which sheep can survive with no winter housing. Every few years there is a great tempest which wipes out thousands, and often bankrupts many sheep farmers. It is at these times that the shepherd, desperate to save his flock, must face the blizzard, armed with dog and spade, and try to find and dig out as many sheep as he can. As Dick Fawcett says:

"Oh, it's very hard work! You have got to probably make a track for them – by pushing the snow to either side with your knees. Well, that's hard enough in itself, and when you have done that you have got to get the sheep to go on the track – you have probably got to drag one and the rest will follow. But on your own then it is a devil of a job, because they won't react when they have been under snow to the dogs. The dog can be pulling the wool off their backs but they are feeling so miserable they just can't be bothered any way.

"So there is a good deal to be said about preparing for it before it happens. We try to never let it happen but, no matter how good you are at gathering, some always manage to get away. Probably the snow is only five or six inches and, when the snow has fallen off, the sheep get hungry so they pop their heads up and start to graze off the patches where the wind has blown the snow away. So you can see them then and you have got to get them down to somewhere and invariably they are between you and several gills. [In the dales, gill means stream.] And that is the hardest part. Getting them through the deep snow you try as much as you can to follow the hills, but you have got to cross the gill somewhere and you can be up to the knees or waist.

"First of all the thing to do is to try to think where the sheep would go in a storm, and then you can go with this dog who scents them. It is no good taking the others – because they would just be hanging around. And you just walk through the drifts until the dog *sets* and that's it: you know there is one in there. And the others that haven't got a dog that sets – they take a big long shaft and they are prodding all the time and I mean it is a laborious job. You have just got to be prodding about there all the time – about every three or four inches – otherwise you are wasting your time.

"And then dig in and get it out. Probably there will be two or three together – it's not often you get one alone.

"Sometimes they are all under; sometimes their heads are out, but they're bad to see and if you didn't have a dog you would never see them. It could just be a stone you see.

"If it's only been under for a day or two and it is in an early stage of pregnancy, then it will just jump up and go where you want it to go. But if it's an old thin sheep, or if it's been under for a couple of weeks, or it's a week or two off lambing, then you will probably have to leave it a day or two until it can stand, and get some feed up to it, and then somehow get it to where you want it – either carry it, or walk it, or sledge it or just whatever you think is the best way.

"We had a bad storm two or three years ago – we had a bad storm last April – and then we had another one last Christmas – so we have had three in three years.

"But you see: even if you only have *one* missing you have got to go and have a look – well if you are worth anything – the same as if you had a hundred missing.

"If there are hundreds up there, they tend to group in a storm – they tend to go to a certain spot. Or the wind just blows them to a wall or something. And if there are a lot of them, they tend to get one on top of the other. You know – it just blows and they pile on top of each other. And you will probably find one and then you dig in and dig, and you will probably find if you are talking in big numbers that the bottom ones are dead. I have a friend up here and he had about twenty or thirty like that – this was in lambing time, and not too far from here – and they were all dead in the bottom of a field. They had all just suffocated. Not the snow – it was just that they had suffocated each other."

As Dick says, only an exceptional dog will find buried sheep. The sheep may be deep down in a drift. The heat of the sheep's body has melted the snow so that she lies in a cavity. Her breath has melted a small hole above her so that she does not suffocate. For a week she can live thus on her condition: the fat that is in her body. Then she begins to pull the wool off her own body and eat it. Then, if she is not found, she dies.

Often, when the snow melts in the spring, its going reveals whole flocks of corpses, and many a mountain farmer finds himself deeply in debt to the bank. Some of them never get out of it, but most survive and live to shepherd again – to build up their flocks of ewes over the years – and quite forget the ordeal that they have been through and forget that, inevitably, they will have to go through it again.

Willie Calvert told me how it was necessary to tread a path over the mountain when you searched for sheep, so that you would see how to get back again, for often the blizzard is blinding. It is easy enough for the stranger to get lost on those North Yorkshire fells in fine weather in the summer time. When sheep are found somehow they have to be got down to the valley. This involves laboriously treading a path for them. Each lost ewe has to have her own path trodden to get her to the main path, and then she has to be driven down the hillside to a safe place. Many sheep are too weak to walk. They must be carried on a man's back, or food must be carried up to them until they recover. Down in the dale the sheep must be fed, but there is hay there for them. If they cannot be got down to the hay, then the hay must be carried up to them. Eat they must or they will die.

A sheep on the high hills lives a very Spartan existence: indeed she is living at the very limit of her endurance. The grasses and shrubs that grow there are sparse and not very nutritious: the coarser species like cotton grass, Yorkshire fog, flying bent (so coarse that it used to be made into brushes), heathers, rush, bilberry and the rest. The sheep have to work hard to make a living at all in such localities. They have little fat in the autumn to keep their bodies going through a bad winter.

Of course sheep farmers are continually breeding for the qualities they find profitable in their sheep: early maturing, good meat production, prolificity, high milk yield for the lambs, high quality fine wool. The type of sheep they aim for is not necessarily one which will withstand wintry conditions at extreme latitudes and altitudes. The sheep was originally a hot-climate, desert animal: it is amazing that it has been adapted by man to live in wet, cold, and snowy conditions.

The Swaledale is now probably the most important sheep in the British Isles. It is a fine looking animal: the ewes have plenty of wool, black heads but white noses, curving horns, but speckled legs; the rams have bold roman noses and no horns.

The ewes in the Dales are crossed with rams of a new breed called the Bluefaced Leicester, described

below. The resulting crosses, called Mules, are avidly bought by farmers from the low country and the south of England. As Thomas Love Peacock pointed out in his imitation ballad:

The Mountain Sheep are sweeter
But the Valley Sheep are fatter
We therefore deemed it meeter
That we carry off the latter.

The mountain sheep have been bred for millennia for their ability to withstand extreme weather conditions. But down in the plains men were able to concentrate on breeding sheep with the qualities they required: early maturing, prolificity, good wool, good meat yield, and so on. Sheep will not fatten on the mountain, nor will they live very long: their teeth wear out. Therefore there has for centuries been a movement of sheep *downhill*. Mountain ewes that have had three lots of lambs must go downhill or they will not thrive. Therefore it has for long been the practice for such ewes to be sold to lowland breeders who, on their lush soft pasture, will obtain several more crops of lambs from them. Of course the lowland men buy these old ewes very cheaply; other-wise they would not bother to buy them.

But the mountain men want to sell their lambs to the plains men too. Very often they can't "grade" their lambs straight off the mothers (that means they cannot sell them fat to the butcher) because they do not grow fast enough, and so they have to sell them to somebody lower down who can fatten them during the autumn or winter. Therefore the ram lambs tend to be sold to farmers from the plains.

The ewe lambs too, most of them, are destined to go the same road. Lowland men will buy mountain ewes from which to breed cross-bred lambs. The mountain men have more ewe lambs than they require for their own flock replacements and so they wish to sell the surplus downhill. But to fetch a good price the ewe lambs must be a cross, or half-bred: pure-bred mountain ewes are too small and wiry, and their offspring slow maturing. So the mountain farmers cross just sufficient of their pure-bred mountain flock with rams of their own breed to provide the replacements for their own pure-bred mountain flock, and they cross the rest with improving rams that will produce a better ewe lamb to sell to the plains men.

Back in the eighteenth century a farmer named Robert Bakewell, of Dishley, by a lifetime of patience, skill and effort, improved the local long-woolled breed of sheep and founded the famous Leicester breed. Some specimens of these were taken to the Scottish border and the Border Leicester was developed, over two hundred years ago, and this has become a great crossing sheep – the ram of all rams. Presumably the infamous Ram of Derby (who had two horns of brass) was of the Derbyshire Gritstone breed, but the Leicester and Border

Leicester became the potent rams of the British Isles.

The Border Leicester was crossed with the Cheviot to produce what was at one time the most common crossbred sheep in England, which became known simply as "the Half-Bred". When I was a boy in England it sometimes seemed as though one saw little else. The Welsh Half-Bred was the progeny of the Border Leicester on the little Welsh Mountain ewe. The Scottish Greyface is the result of the Border Leicester on the Scots Blackface. And so have all the breeds been created.

A new breed of sheep has evolved from the Border Leicester in the last fifty years, and that is the Bluefaced Leicester. This has been bred for the sole purpose of crossing with mountain breeds for the production of ewes to sell to the lowlands. It is the Bluefaced Leicester that people like Dick Fawcett breed simply to use as rams to put on their own local mountain sheep, in Dick's case the Swaledale. The Bluefaced Leicester-Swaledale cross has become known as the Mule in most parts of England and is rapidly becoming the most popular sheep.

But the story does not stop here. When Dick Fawcett and his brethren of the hills sell the Mules downhill they are bought by lowland farmers who

Swaledale

Derbyshire Gritstone

Blue-faced Leicester

Scotch Black face

Welsh Mountain

Mule Lambs

cross them with yet another breed of ram. As we shall find when we reach the fertile plains of southern England, the farmers there go north every year to buy replacements for their Mule breeding flocks, but they do not buy Mule rams. They use rams of southern lowland origin – from one or other of the Down breeds, of which there are many to choose from. These Down breeds would not survive on the Pennines or the Cheviots for one winter, but they impart their good commercial qualities on their progeny by the mountain-bred Mules.

There are no less than forty-eight recognised breeds of sheep in Great Britain, which is probably as many as could be found in the rest of the world. Pedigree sheep from many of these breeds have gone all over the world to improve local varieties. Their only rival is the Spanish Merino with which, for wool production, they cannot compete. British sheep have been bred for mutton, not wool. Wherever good mutton and lamb meat is sought after – which means nigh-on every country in the world – you will find the blood of some of those forty-eight breeds of sheep.

One evening Dick Fawcett told me about his shepherd's year.

"In October we decide how many ewes we want

to replace. We have a flock of six hundred pure Swaledale ewes – we probably want between a hundred and a hundred and fifty replacements."

In the very high mountain flocks, as we have seen, ewes are culled or draughted after they have had about three lambings; their teeth are no longer adequate for eating the rough, mountain pasture. But such ewes will give another three or four crops of lambs, quite happily, down at lower altitudes. So they are draughted out of the mountain flock and sold to farmers from kinder lands. The Fawcetts own quite a lot of valley land and so are able to make use of these animals themselves.

"So we go to auction mart at Hawes and buy maybe not the very, very best because they would invariably be bought up by somebody who wants to buy a few and breed tups himself. [Tups means rams. Such a purchaser would be wanting to breed very high class tups for sale to other farmers at high prices.] But we will pay maybe fifty to sixty pounds a piece for these ewes, which are four-shear – which means they have had three crops, three previous crops, of lambs. And they are always uncrossed, which means they have never been crossed with anything else – they have always been tupped by a Swaledale. Because there is this belief in Swaledale – it is only a lot of baloney – but a lot of men believe that, if a sheep has been tupped by a Bluefaced Leicester or a Teeswater or anything, it is never fit to breed Swaledales again. They think it is contaminated, that its future offspring will never be pure. They still believe it to this day, a lot of them."

Although the Fawcetts keep their sheep for most of the year on the high fells, they also, as we have seen, own a considerable slice of land in the lush valley bottom. It is therefore possible for them to treat their lambing ewes with far more consideration than can those whom we might call true mountain farmers, the ones who are confined to the mountain only. Hence they can buy older mountain ewes as replacements for their flock, and they can flush their ewes hard before tupping so as to achieve plenty of twins. Farmers with mountain pasture only do not welcome too many twins (the ravens and foxes eat them), and they cannot bring the ewes down to lamb in the valley. These must lamb high up on the rough mountainside, and take their chance, and if blizzards come while they are lambing they are lucky to survive themselves, let alone save their lambs. Dick Fawcett's ewes are considerably more fortunate.

"We buy these ewes in October and we fetch them home; and we fetch the rest of the ewes off the fell about a fortnight before tupping, which is about the fifth of November, fetch them down into the fields to flush them. [Flushing is the process of taking ewes off poor grazing and putting them on better pasture shortly before, or when, the tupping takes

place. Tupping means serving by the rams. If ewes are flushed they take the ram more quickly and readily, being in rising condition, and thus tend to lamb more at the same time, and to carry more lambs in their wombs.] And then we sort them out into lots of maybe about thirty up to a hundred and twenty. You give them each a different pasture and each tup has a pasture to himself. We are trying to breed a particular type of mule lamb you see. There's a very good premium for the right type of mule gimmer lamb. [A gimmer lamb is a female lamb, and a gimmer hogg is a female between weaning and shearing. After the first shearing at a little over a year old she is a shearling gimmer.] And it is up to us to try to breed the best type of mule lamb we can. We have a very good bunch of tups which we have bred mostly ourselves over the years, and we are selling a lamb in the market as high as anybody else. The top price in the market is making about seventy, eighty, ninety; but on average they are making about forty, fifty pounds apiece.

"Our tups are a Bluefaced Leicester. We breed them ourselves – we have just a handful of about four or five ewes, pure-bred. To breed our own tups we either use our own or we'll ask someone who has a particularly good tup and we will take a ewe to him, and he will take one of ours. There is no payment involved – it's just a case of trying to help one another.

"If we have a particular good tup, he will get a hundred and twenty ewes to serve. If he is a tup lamb, first season, he will get twenty-five, just in case he's no good.

"And they stay with the tups for as long as it takes and until we're sure they're in-lamb. [In-lamb means with a lamb inside them.]

"At the end of each week we mark them as first-weekers, second-weekers, third-weekers, and we mark each ewe according to the tup she was with, so that we know when she is due to lamb and which and what the sire is."

I must here explain that this tupping takes place, not on the fell or mountain, but down in the lush grassy bottom of Wensleydale. The bottom of the dale, like all the other Yorkshire dales, is enclosed in fields by limestone walls. The good, improved grass in these fields is grazed by dairy cows – which provide the milk for Wensleydale cheese among other things – but is preserved religiously for the sheep at certain critical times of the year: for flushing, tupping and lambing. In between times the sheep are pushed back up on to the fells, to preserve the lowland grass.

"We turn them out on the fell about the first week in December. It's an enclosed fell of about eleven hundred acres which we share with my uncle. We have six hundred sheep and he has coming on for five hundred.

"They stay up there as long as the winter will permit. It isn't a particularly easy fell in winter because the best land is up at the top and if the weather is particularly bad it drives the sheep into the bottom. So once it gets into January we start putting Rumevite blocks [highly nutritious cubes, which the sheep lick] about, and feeding a bit of hay and a bit of dried sugar beet pulp and we carry on doing that as long as the weather is fairly reasonable. But, if it becomes wild and wintery, we fetch them home. But we think carefully about fetching them home because it's four miles, and once they are there you think twice about putting them back.

"When we take them back up to the fell we walk them down right through Hawes along the road. It is pretty quiet in winter, and if it's summer we just go very early in the morning. We try to leave them up on the fell for as long as possible.

"We have two hundred acres down here in the valley bottom and a hundred and twenty acres of rough rocky pasture up behind. Which is very good for saving the sheep when it is wet and puddly down here. We try to keep them off this land as long as we can, but we lamb them down here in a pasture which is particularly sheltered and behind a big wood."

But lambing comes early in the spring and it is then that snow can come too.

"Well, we had one big snowfall during lambing time last April. We didn't suffer too badly, but the higher men suffered badly in parts – they lost hundreds of sheep. It was quite a miracle was that, because we would have been three-quarters of the way through lambing. And there were lambs born that day and there were lots born the previous day and we didn't lose a lamb. They all survived, and we had some of them under drifts for twenty-four hours. We had five hoggs [young sheep] up there of last year's lambs and we had them on the fell and we lost two out of five. The others were just lucky to be in a safe place.

"Well, if it's really bad there is nothing you can do. Because you can't see anywhere, and if it is really wild the sheep are as we say 'darking' – that means sheltering – and they couldn't hear you whistling because of the driving winds, so there is no point doing anything until the weather has settled down.

"But, if you can get to them before it happens, you have got to get them down to the field where it's more sheltered. So, if it becomes a really bad blizzard, at least we know they are in that parti-cular area and we have only got fifteen acres to look around as opposed to eleven hundred."

I tried to imagine roaming about on that vast expanse of wild moorland in a blinding blizzard trying to search for buried sheep.

"Usually, if we get them down, they are safe. It doesn't often happen that when you get them down to that altitude there is much snow, but it has happened. It happened these last couple of years but that was a bit of a freak. During the winter you have really got to be on the ball. You have got to look at the barometer, look at the sky, and listen to the weather forecasts. And if you think it is going to snow, then you go up and dog in and gather down."

But to return to the intricacies of lambing. The sheep, like the human, is intended by nature to give birth to one offspring at a time, and occasionally two. The incidence of twins in sheep is increased by flushing – putting them on to better pasture just before and during tupping. By selective breeding, mankind has caused domestic sheep to have a larger than natural number of twins and, more and more in these days of ever-increasing human greed, triplets and quads. Quins are by no means unknown among some continental breeds of high-producing sheep.

But in the hills and mountains twins can be an

embarrassment and have to be given special treatment:

"The singles stay out. We ship them out twice a day, as they are born, to clean fields and pastures. The twins we fetch home, and we put them into straw-bale pens in a shed for twenty-four hours to get them mothered, cleaned up and dried, and then we take them out to the fields and keep moving

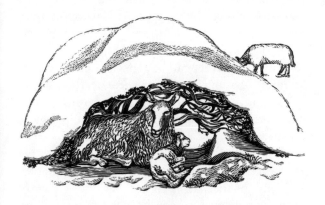

them away and moving them away [away from the farm to encourage independence]. All this time we are feeding them up to three-quarters of a pound of sheep nuts and hay *ad lib*. [Feeding *ad lib* means giving an animal as much as it demands.]"

The routine of lambing on the stone-walled

pastures in Wensleydale goes like this:

"I get up at dawn, which is no earlier than half past five, and I go round the lambing sheep in case there's something lambing incorrectly and needs assistance. And that might well be the case. There could have been some lambs born on a frosty night, born on the freezing hard ground and just lying there – just ticking that's all. So we rush those home and we always put them in a sink of warm water to get them living again. Then we dry them and put them under a lamp and give them a protein injection, which works wonders – protein and vitamins. And they can be perfectly normal within an hour – having been lying there nearly stiff.

"So you go round the lambers and, once you are convinced they are all in order, then you go round the next youngest lambs. Just looking to see if everything is all right: like most living things, they are most prone to problems when they are babies.

"And the sheep have all sorts of things go wrong with them as well, so you have got to have a very very keen eye. It is always easier in a nice quiet lambing time than it is if the winds are blowing like mad and there are sheep in every corner, and snow everywhere, and you can't tell whether a sheep is all right or not because they're not going to come running up to you.

"We always lamb in the pasture along the road there, because it is very sheltered there behind the big wood and it is always very easy to get to it, always sound ground which you can drive across without making a mess. The sheep can lamb there all through lambing time without getting it at all puddley.

"I always have a small trailer at the back of the Land Rover, which will carry about eight sheep, and in the back I have some boxes for the lambs, for about seven pairs of twins or whatever it is I want to bring home.

"Back at home I fasten the mother up somewhere until the lamb is fit to go back to her. We make a lot of pens out of straw bales and we have several buildings that we make use of.

"If the sick lamb is a twin I take them both off the ewe, because, if you just leave her with the healthy one, by the time you give her back the sick one she's forgotten it and won't have it back.

"If you find a dead single lamb, you just skin it and put the skin over one of a pair you want to foster."

It is very difficult to get a ewe to accept another ewe's lamb. But if a ewe has a dead single one, and if her udder is sound, it is always good policy to foster one of a pair of twins on her, for no ewe can bring up twins as well as she does a single lamb. The age-old trick is to skin the dead lamb and put the skin on the lamb to be fostered like a jacket. After a few days it can be removed. There are other methods though, as will be explained.

"Maybe a sheep hasn't got enough milk in her teats – she has maybe had mastitis or something – well then we take one lamb off her. I often have a rucksack on my back and I put one of these lambs in the rucksack. Then, if I come to a lambing ewe and she is a fit sheep and is only having one lamb, I lamb her and then I cover the lamb on my back with fluid – absolutely soak it – and hope that she will take it.

"You can get little pens where you can foster them on – they have a space where they put their heads – and they seem to be very successful. We don't use them, but that's because we don't do an awful lot of it I suppose. But the down-country men do when they have a big crop – they often have triplets and things like that.

"Then we go – well you're almost at it all day. We go round after breakfast, at about nine, then we go round about one, then about four and then about seven, just before dark.

"And they stay down here until the grass is started coming on the fell – which is usually about the tenth of May. Then we start trucking the singles back up again – just the singles. What we do: we put the lambs in the truck and the mothers in the trailer, about twenty-five at a time. And then we take the load through Hawes and unload them, and one of us starts walking them up the road. Then I go back for another load and I take that load to wherever the others have got to and let them out there. And we carry on like that. We might take about seven or eight loads a day.

"And the twins stay down here all summer, and after about a fortnight old they are all marked and castrated and tailed, twins and singles."

When I was a boy working for a lambing season on the Romney Marsh, the old looker I worked under used to castrate the lambs by cutting off the end of the scrotum and drawing the "stones", as he called the testicles, out with his teeth. I could never bring myself to do it. Nowadays such practices are obviated by the use of an instrument called an Elastrator. This slips a tight rubber band around the scrotum, which cuts off the blood supply, and the scrotum, and testis, painlessly (we hope) fall off. The same is done to the tail.

You may wonder why it is necessary to remove the tails from animals which are obviously meant by nature to have them. Well, true mountain sheep which live up on the high mountains all their lives, or sheep that live in the desert, can carry their tails throughout life and enjoy them. But sheep that are fed on lush pasture, with supplements of concentrated food, will "scour" (have diarrhoea) and foul their tails, which will then attract that curse of domestic sheep, the sheep fly, the maggots of which will eat the sheep alive.

Sheep that survive on the high hills, though they may suffer from the weather and from predators, are in many ways healthier and not so prone to disease as the lowland breeds. The Fawcetts, like all sheep masters, have to take precautions.

"The Fell lambs are dipped once, that is today when we have them down from the fell and we clip the hoggs at the same time. We dose them for stomach worms. The lambs down home we dose when they are about six weeks old, and then we dose them about every month after that; because,

with them being far thicker on the ground and with ground that has had sheep on it all the year round, you know it is the only way."

Dick is referring to the build-up of parasitic worms on such land, which does not occur on the much more sparsely grazed fell.

"We vaccinate for orf in early June and we have had one or two really bad goes with orf. It drove us bonkers all year – every lamb had it on feet and mouth."

Orf, which vets call Contagious Pustular Dermatitis, is one of the many diseases of domestic sheep. I have a book which lists *ninety-five* such diseases, all of them found in the British Isles. I cannot believe that the wild mountain sheep on the Atlas Mountains suffer from anything like the number of diseases on this doleful list. Old sheep farmers say: "As soon as a lamb is born it starts trying to die!" Another such saying is: "The worst enemy of a sheep is another sheep!" It often seems as if both of these are true. The fact is that the domestic sheep has been bred selectively to such an extent that it is very far from its original wild condition. Health and robustness have been lost in the search for unnaturally high prolificity, quick maturing, heavy and fine fleece, and a good meat carcase. Grazing animals in the wild do not suffer unduly

from worm infestation: the small number of worms which they do get have been called "healthy worms in healthy animals". On the other hand sheep confined to small areas of land keep picking up the eggs and larvae of their own parasites. Hence the modern sheep farmer has to be constantly on the look-out for disease.

I wondered who was available to do all the vast

amount of work Dick had been describing, for the farm has a large herd of milking cows as well as the sheep.

"There's my father who is at retiring age and there are two men and myself. My father is also manager and chairman of Hawes Auction Mart which is quite a big responsibility. So all his time is spent there – particularly at the back end of the year. I do most of the sheep work and my father does a bit. I'm an only child."

Dick went on to describe the rest of the year's work. In mid-July they have to shear the ewes with the twin lambs which have stayed down in the dale. Then these ewes have to have their feet trimmed and

be put through a foot bath. Sheep up on the fell don't get footrot – a very common and debilitating disease of sheep – but down on the lush pastures they do. Even the least observant townsman out for a drive in the country must have noticed sheep "praying" – down on their fore-knees while they are grazing. This is because their rotting feet are too painful for them to stand on. Good shepherds such as the Fawcetts do not allow footrot in their flocks. It is said that the remedy for footrot is "change your shepherd".

Then the main job of the year begins: sorting through all the sheep, draughting out sheep for sale, and "making up" the flock for next year.

"We go on the fell late July early August; it depends on the hay time – hay time can go on until late August and it generally does. We make seventy acres of hay. We get about sixty inches of rainfall here – six miles up the road you get a hundred.

"The lamb sales start in about the last week in July. We often sell ours at the first sale. There may be a second sale as well. But our twins are only small – no age really. And then about the middle of August we fetch the lambs home. And then we start selling them: this is the wethers – the gimmers we don't sell until October. [...Wethers are castrated males; gimmers are females.] We sell the wethers as stores – we find they don't fatten well. [Stores means unfat animals. These are sold off for somebody else to fatten and then sell for meat.]

"They sell between two thousand and three thousand of our store lambs at Hawes and they go mostly to places like the Vale of York, but they go all over the place: to the people down on the low-land. The big lambs will be fattened within a few weeks but the smaller lambs will be kept on until next Easter, when they will be worth quite a lot of money. Mostly they're just scavenging – on sugar-beet tops and things like that – just tidying up another farm.

"The gimmers are worth more than twenty pounds apiece more than the wethers. This is why we place such importance in getting good tups you see. They will all go for breeding, the gimmers, practically every one. The very best go to Cumbria really, there there are a lot of very proud men who like to have good sheep and they buy the tup parent too. But the rest go absolutely everywhere – a lot go to Hampshire, and to the Midlands, and once again the Vale of York – but they go all over the country.

"Farmers come up more and more than they used to. There used to be one or two men come and buy twelve, fifteen thousand ewes and then take them down and the farmers would buy them from them, but now most of them come up themselves. There are about three hundred buyers and they have a two-day sale of about thirty thousand; three weeks after that they have a one-day sale of about fifteen thousand; and then three weeks after that they have another sale of about five thousand. They are all Mules [i.e. Bluefaced Leicester/Swaledale cross]. We don't keep any at all.

"Well, once we have weaned the twins we take those ewes to the fell again. And all the ewes stay up there and we don't take them down again until tupping time either to flush or sell.

"We sell the old ewes at the auction mart. We'll probably draw out about twenty of the 'correct' ones – that is correct udders and mouths – and they'll be taken down country and somebody will probably get another crop out of them. The rest of them – the Pakistanis eat most of them. Most of the old ewes go to them and they get curried – the old cast rams too about Christmas time, when they are all worn out and tired: they come up and buy them and they must be tough stuff. Sometimes the Pakistanis come up themselves, not very often though, usually dealers come to buy for them."

The advent of the Pakistanis in England, and the Turkish "guest workers" in West Germany, has brought a solution to an old problem – what to do with the meat of old, worn-out ewes and rams. Many farmers have been puzzled at the fact that, nowadays, you can never buy mutton at the butchers: it's, "No we only sell lamb!" So where do all the old sheep go? Well, they used to go to the poorer parts of the industrial cities. Now they end up in Pakistani and Indian restaurants.

The big sheep markets in the north of England "at the back end of the year" are events of immense importance in the sheep men's lives. Up come the farmers and dealers from the south to compete fiercely for the Mule gimmers which are the real product of these mountains. Some also come for store lambs to take them down to fatten on better land and sell at a profit, but most come to replenish their own breeding stocks in the lowlands. They will put heavy Downland rams on them, flush them heavily, feed them well, lamb them indoors as often as not nowadays, and expect to get big crops of fast-maturing lambs for the luxury fat-lamb trade. But we will follow the fortunes of Dick Fawcett's mule gimmers as they go downhill in due course.

Meanwhile the breeders of pure-blooded Swaledales – to be the mothers of the Mules – have a little gamble of their own.

Pure Swaledale rams, or tups, change hands for thousands of pounds. Five or six thousand is often paid for a splendid animal, with a noble head, great curved horns, and fine conformation. I asked Willy Calvert how it is that fell farmers can afford to pay so much money for a ram.

"It's like this. Now I have made a lot of money this back-end now tax-man is going to go away with the lot so let's see about a gay good tup: we'll spend a lot on a tup. And that's – that's how it escalates you see. Well it were worth over a hundred pounds – a hundred was worth a lot of money in them days you see. Now it's thousands."

The whole thing reminds me somewhat of football transfers. A club will pay hundreds of thousands of pounds to another club for a player. Where did the money come from? Well, of course, from another club who paid *that* club hundreds of thousands of pounds.

The Swaledale enthusiasts buy rams for thousands – but then they sell their own rams for thousands, so the money simply circulates around. But it adds to the excitement of the ram sales: the crowd round the ring stands with bated breath as the bidding goes up and up – and, of course, the lucky purchaser obtains by his apparent extravagance great publicity in the world of sheep men.

The Swaledale is becoming pre-eminent in England. It is on a rising star. The Bluefaced Leicester/Swaledale cross, or mule, is the most sought-after ewe for crossing with the meatier Downland rams. Willy Calvert, who lives in Swaledale itself, is nothing if not sanguine:

"Of the Hill breeds is Swaledale and we are ousting the other Black Faces and other Hill breeds on that. And new gimmers are running – that go off these sheep – are running the other breeds out down-country – all the fat breeds and Down breeds and that. These sheep can breed a lamb that's ready and better than anything owt could you see.

"It will never run them out altogether, but there is still room for everything they can produce yet. Some people think it will come when we can out-do the market eventually but it is not likely yet. The market is expanding at the expense of other breeds you see. And you can only say that there is a terrific future for the Swaledale sheep that go off these hills.

"Yes – it is in other breeds – like the Scots Blackface – the tups were making likes of five thousand and *ten* thousand long before we were. And when we were down to one or two thousand for top tups they were up there and they probably still are in their pure breeds and there will still always be a lot of pure breeds in Scotland, and England as well you know. It will never cut it out. But a lot of Scotland men are buying our tups and crossing their breed with it. They are Scottish Blackface with a Swaledale tup and they get a ewe that shows a bit of Scotch but they get a bit more life and action and moving onto't moor. And eventually if they keep crossing they will come up with a ewe that will be like ours. Whereas with their Blackface ewe they might make thirty pound a head – if they are three-quarters Swaledale and showing nearly all Swaledale they will make nearly sixty pound a head – well that's what they're after isn't it?"

I don't know what a Scottish Blackface breeder would think about *that*! But so the battle of the breeds goes on, arousing passions and emotions of which people who are not sheep people can have no inkling.

Dick Fawcett is a bit more charitable to the Blackface, the Swaledale's chief competitor:

"I suppose the Blackface ewe will do better in some areas than the Swaledale. It is true that some sheep are suited to some land more than others. Some people won't have the Mule because it doesn't grow much wool, particularly under its belly. Well, if it has got a bare belly during winter and it is on wet clay ground, then obviously that is going to affect it. Whereas a Masham [the result of a Teeswater ram on a Dales Bred, Swaledale, or Scottish Blackface ewe] is a big woolly thing and it can sit there all day and not use any energy keeping warm because it is already as warm as toast. So it is probably suited to some areas rather than others – well, that is what some people tell you."

But in Swaledale and Wensleydale the Mule is king, and hundreds of buyers from the south and the northern low country come there every year to meet thousands of sheep.

"It is a sale with a great atmosphere you know and the ring is always packed out. And a lot of auctioneers do a good deal of bantering you know! You know – they make a big thing out of it. And it's great to be a part of it – I love buying there. You have got to be careful because there is a hell of a

trade for the best ones – it is always the same with any commodity, especially livestock. If it is a good one, there is always a hell of a demand for it; if it is second rate, you can always put a value. But good stuff will make a good price and it might even make a better price than you thought it would! Yes – it is very exciting! I mean I really look forward to it. I know all the sellers there and I go round and pull the sheep to bits and say I'm not buying this year, what a sorry looking pack – and then go and buy them you know. Yes we have some good fun.

"It is the same with the tup sales. That particular week, they call it 'Swaledale Week'. The registered sheep are sold on the Tuesday, and then the ram lambs are sold on the Wednesday, and then the shearling lambs are sold on the following two days; so it is a full week. And the pubs are full every night and everybody is arguing and shouting like mad and giving you an awful lot to drink and it is 'tups – tups – tups' – nothing else is ever discussed except sheep and it's a good laugh.

"A lot of the buyers come and stay overnight for the mule gimmer sales – they make a holiday of it a lot of them. And we get to know them, and we go and meet them in the pub and prime them for the next day – fill them up with propaganda and drink. Not too much though – not sufficient to make them too ill to turn up the following day! Yes it's like an annual reunion. They like to meet the characters – like in all remote places you will find characters and there are still a few around here. Buy 'em enough drink and they will keep talking all night for you.

"There are plenty of other markets as well but I think I am probably right in saying Hawes is probably the best now. My father, I think, is as much reason as any."

II

LOWLAND SHEPHERDS: MIDWIVES AND NURSEMAIDS

THE FAMOUS Mules are bought and sold, and down they go to the lush lowlands, in their thousands, to join there Welsh Mules, Scottish Mules, Welsh Mashams, Scottish Mashams, and various assorted crossbred sheep out of a dozen or more Hill and Mountain breeds of sheep. "First cross vigour" is a well-known factor among all stock breeding men, and these first crosses, with the vigour of their mixed parentage and also the vigour and hardiness that comes with having subsisted for generations in extreme conditions – will flourish well in the fertile lowlands; and the additional crossing that will happen there with the heavy and well-fleshed Downland sheep will increase the value of their progeny. So, we will follow a bunch of our Blue-faced Leicester/ Swaledale Mules down to the south of England.

The shepherds of the Scottish highlands and the upland regions of England and Wales are fine and hardy men indeed, with unrivalled knowledge of their craft; but it is the south of England shepherds who know most about intimately caring for sheep.

In the mountain areas the sheep, perforce, have to get on with their lives by themselves for a great deal of the time – for virtually all the time, in the case of the true mountain sheep. When you have thousands of sheep, spread over miles of rough mountain – apart from dipping them and giving them the odd injection – you have to let them get on with it, even at lambing time. It is in the lowland areas of England that the true caring shepherding is practised – where every sheep is always under the shepherd's eye and every lamb is lovingly brought forth.

Tom Butcher, whose story we will hear presently, is the traditional lowland shepherd *par excellence*.

Ian Williams is a new kind of shepherd: not of farming stock at all, well educated academically,

and then at an art college, where he met his talented wife Sue (the author incidentally of a delightful little book for children, *Lambing at Sheepfold Farm*). Ian has taken to shepherding first to make a living in the sort of countryside he prefers – the sweeping downlands of the Hampshire-Dorset border – and secondly, quite simply because he loves being a shepherd.

In a very few years he has mastered what is a very complex and difficult art – an art that has its origins back in the infancy of mankind. Ian still paints, and has exhibitions and sells paintings, and he will talk about painting too, but it is when he talks of shepherding that his eyes really light up and he warms to the subject as he warms to no other. But his talk is highly technical, because this is a highly technical subject. Most farming people will not bother to talk to non-farmers about their profession, because they know there will be blank incomprehension. If a city person walked into one of the pubs at Hawes, full of sheep farmers at the time of the auctions, he would not understand a single word of what anybody was talking about. It would seem like a foreign language to him.

The farm on which Ian Williams was shepherd when I last saw him was a huge one – much larger than any one man ought to own – and he was just beginning the task of lambing down seventeen hundred ewes! If you can imagine a maternity hospital with seventeen hundred mothers-to-be in it, one surgeon, one night-nurse and the occasional help of the surgeon's wife you get something near this situation.

You may object that seventeen hundred lambing ewes are unlikely to give as much trouble as seventeen hundred women. But these were not ordinary, natural ewes. They had been steamed up on high feeding at flushing time, had been fed high ever since, and many were half Friesian which means that they were quite likely to have triplets and quads. Even the Mules that we have followed down from the Pennine Hills, because of their high feeding, would have many twins and triplets. Also the very high feeding, which they had had all winter in order to stop them reabsorbing their lambs (a thing undernourished ewes are quite capable of doing) and to ensure that they would have plenty of milk for the large numbers of lambs which it was hoped they had in their bellies, causes much trouble at lambing time.

It is very difficult to strike the balance between under-nourishment and over-fatness. This entails very close and highly intelligent management. But no matter what the management, people keeping

sheep in this way can be sure of a high proportion of births needing assistance, a great deal of work and worry about the lambs afterwards. Multiple births in sheep always lead to mis-mothering, lambs being deserted, lambs having to be foisted on ewes that are not their mothers, and a certain amount of disease. Mastitis is common and so is prolapse: the extrusion of the vagina or even the whole womb. I have seen Ian catch a ewe which had her vagina hanging out the back like an inverted balloon, fling her on her side with a twist of her head and a pull on the shoulder, replace the vagina and sew the orifice shut with a carpet needle. Within a minute of being caught the ewe was on her feet happily eating hay.

Everything in such a lambing shed has to be done at top speed: there is just no time to linger about. Just as we were leaving the lambing shed one night we saw a ewe in the first stage of labour. The night-shift would not be there for a few hours. This ewe would have to be left to get on with it herself, or we would have to stay with her (and we were hungry) or she would have to be delivered forthwith.

I had been taught fifty years before by the old looker on the Romney Marsh that you should never interfere until the ewe has been in labour for at least an hour. Always let her do it herself if she can. But this is not the way of modern shepherding.

Ian laid the ewe down, pulled on a rubber glove and oiled it, inserted his hand and pulled out a lamb – inserted his hand again and pulled out another lamb – felt her tummy, said: "There's another one inside!" – inserted his hand and pulled out another one. We carried the lambs – the ewe bleating behind – to a pen, put all four animals in, washed our hands under the tap and were walking away within, I suppose, five minutes. And next morning mother and children were all doing well. One of the triplets, of course, had to be foisted on another ewe. All was well.

I describe these incidents to illustrate the extreme pressure that shepherds are under in such large and intensive sheep enterprises and the degree of expertise that it is necessary for them to have. Now I must set the scene.

The country on the Dorset-Hampshire border, south of Salisbury Plain, north of the New Forest, consists of sweeping open downland, with occasional woodlands retained for the nurture of pheasants. There are some genuine farmers of old yeoman stock, but most of the land is owned in huge holdings by people who have made their millions in such places as the City of London. It is the sort of situation foreseen a hundred and sixty odd years ago by old Cobbett, as he rode about these very downlands on a horse and wrote his descriptions in his *Rural Rides*.

So, as you stand outside the back window of Ian's thatched cottage, you look over a great sweeping landscape with few hedges or fences, hardly any dwellings of mankind, just the occasional mansion here and there and very few villages. It is, in one of the most highly populated countries of the world, not a hundred miles from its capital, an unpeopled countryside.

The small farmer – the smallholder – has no place here at all. There is no little chink or cranny into which he can fit. Only the large landowner, generally an absentee except at weekends, has a place here. Agriculture has to be arranged so as to employ very very few people – because there *are* very few people. The people who closely populated this land in Cobbett's day all went to the towns, or "the Colonies".

There being very few people to look after them, there are therefore very few animals. (Horses excepted – there are plenty of these. Not for work, of course, but for rich people to ride about on.)

However, there are still, in places, some sheep. These chalklands used to be covered with sheep.

In 1800 a Mr Luccock estimated that there were over 700,000 sheep in Wiltshire alone! Now you could probably knock two noughts off that. But a few of the huge arable farms still do keep sheep – good deeds in a naughty world – because, on this light land, the sheep is *The Golden Hoof*. The cloven hoof of the sheep, plus its urine and manure, enriches and improves light land as nothing else can. A remarkable thing about sheep is that adding them to a farming enterprise does not necessitate the reduction or elimination of any of the other farming activities. If you keep cows or beef cattle, you can add one sheep per head of cattle without robbing the cattle, for the sheep graze closer than the cattle and eat grass that the latter cannot get at. They nibble round the droppings of cattle at the lush grass there that the cattle themselves will not touch, they "clean up" pasture after cattle and they do nothing but good. If they are kept on an arable

farm, it is true that special crops have to be grown for them, but it is also true that they immeasurably improve the land. It is true also that in these days when all fertility comes from a bag (in the form of artificial fertilizers and biocidal sprays) many arable farmers manage to get on without animals at all. But, as the oil gets lower in the oil-fields, the day of "the bag" may soon be over. Then farmers will want to turn back to sheep – and where will the shepherds come from?

Standing out on the wind-swept downland, with a tarred road leading to it, stands the lambing shed in which Ian does his lambing. It is a huge shed. It seemed to me as big as a football pitch. Just a minimal building – a huge roof on poles, open on some of its sides and closed in on others. The ground in it was covered with straw. Arable farmers with no animals burn their straw nowadays: at least the straw on this farm was being put to good use. First it was giving comfort to the sheep; secondly, rotted by their dung, it would give fertility – real lasting fertility – to the land.

Inside, the building is cut up with hurdle fences. There are several large sub-divisions and many small coops, or pens, in which individual ewes with their lambs may be confined. In the middle is a wash basin with hot water provided by a small geyser, and a table containing various surgical supplies: syringes, needles, both hypodermic and for stitching, rubber gloves, antiseptic sprays and ointments, oils and disinfectants.

The whole seventeen hundred ewes that have got to lamb do not go into this shed at the same time because, as Ian says:

"The rams have harness on, so that when the ewes are tupped it leaves a mark – it's a wax crayon – on the backside of the ewe so that we know when they are going to lamb. We use a blue raddle to begin with – we divide our numbers into three, as we can only get a certain number of ewes into the barn. We try to aim for about five hundred and fifty, so when we've got six hundred blue we just change the raddle. We don't do it by the day – we do it by numbers. There's just a rough count of blue bottoms in the field and then we change to red, and then when we've got five – six – hundred of those we change it again. And they run split in January, when we take the rams out, split into their colour groups and they are fed accordingly.

"And, of course, the concentrate feeding which in our case is rolled oats, sugar beet and fish meal – that usually starts six to eight weeks before lambing and is built up to about a pound and a half at lambing.

"They are lambing in April and you hope the grass is beginning to shoot away, but because it's so heavily sugared, as they call it, on our farm the magnesium is locked up. [Sugared means dosed with nitrogenous artificial fertilizer.] And like cattle, lambing ewes need magnesium and they can't store it, so we have to feed cake after lambing to get magnesium into them. So, even though we have a foot of grass, people wonder what the hell we are doing – we are going out there with bags to them! It's really to get mag into them. Any ewe that doesn't come to the trough is given a private dollop. Last year this went right on till May, which is expensive business – but then you lose a ewe you see. We lost twenty ewes last year – we might have lost a hundred if we hadn't fed, so it's an expensive business."

Hypomagnesaemia, or Tetany, attacks grazing animals which are yielding much milk on grass, when the grass has been "sugared", as Ian calls it, with large doses of fixed nitrogen. Milk Fever attacks heavy yielding ewes – or cows – because their bodies cannot get calcium quickly enough. Twin Lamb Disease is thought to be caused by a shortage of sugar in the liver of very high-yielding ewes, so that the ewe cannot use her stored body fat properly. All these diseases affect high-yielding ewes and they can all kill, very often quite quickly. These are just a very few of the killing diseases that the modern shepherd, caring for very high-yielding sheep, must look out for. They are the problems caused by keeping sheep in conditions that get further and further away from the conditions for which sheep were evolved. The modern shepherd knows as much as any vet about these things, and is equipped to deal with them. Ian would give a massive injection of calcium borogluconate to any ewe with milk fever, and intravenous injections of glucose, or high protein, for other conditions.

Let us follow Ian's work throughout the year.

"We wean the lambs in September and we start then thinking about next year's lambing. We've still got quite a lot of lambs on the farm then – we may still have a thousand lambs on the farm. Half of them will probably be finished fat, but all the lambs are off the farm by November, apart from the ewe lambs. The ones we can't fatten are sold as store lambs – for other people to fatten: dairy farmers or specialized fatteners. Those people will put them on various crops – a succession of crops, probably autumn rapes or stubble turnips, which is very common now, of course, usually put in behind winter barley. [Barley sown in the autumn – 'winter barley' – ripens soon enough to enable a 'catch crop', as it is called, of turnips to be sown into the stubble, and this crop can be fed off to fatten bought-in store sheep.] Probably when that peters out they may run round on some kale and swedes and people will probably have to cake feed them a bit – either whole barley or rolled oats or a proper protein sheep nut.

"But by then we're really thinking of next lamb-ing. We allow ourselves eight weeks to turn the ewes round, and that's to cull them [cull means pull out old or unsound ewes to sell], and sort them out for condition – which means you sort the thin from the fat. The very fat ones are literally starved on the old grass, which is due to be ploughed up so the amount of grass there is almost nothing. It is the traditional idea, and it is probably right, that if they are too fat their conception rate drops. So the idea is to let them down, and then flush them, which is

Suffolk

the traditional word for building up their condition rapidly, and hold it there just before and after tupping, and then after tupping for a month; and this is a very tricky thing to do. You want the condition to come down and then bring it up.

"But for the ewes that are in poor condition in September – you might have young ewes that have milked hard, or old ewes that have had a rough time – they are drenched [dosed with worm medicine] and run over the new leys that we planted behind the winter wheat or winter barley in the summer and have probably got ten or twelve weeks growth in them in the autumn. [A ley is a plant of new grass and clover.]

"This year we had about three or four hundred, and these are the poorest conditioned ewes, and if need be (if they were very poor) we would have to

Hampshire Down

cake feed them as well, but we only do that in a drought year.

"By keeping the ewes hard in the summer, of course, you've been resting the fields and they've grown in a good year – they've grown very well in September and October. So far it's worked. We don't grow any special crops for the autumn flush-ing and we probably should do. We grow our kale and swedes – we usually start feeding them about December and January time to carry us up to lambing.

"So our tasks from weaning the lambs would be: culling the ewes; sorting them out in their condi-tion; turning every single ewe up [upsidedown] and foot-paring them – and any chronic footrot or proud flesh on the end of the foot we use hot irons to try and cauterise it, and any ewe with strawberry footrot or any infection is marked and kept apart from the others. All the ewes are run through a very long footbath – forty feet long – and stopped there.

"The rams meanwhile are also being foot-pared at least once a month and we've got forty-five rams and that's a day's job doing those really. Heavy old boys! They are fed cake from six weeks before tupping to bring them up into condition. And, if we've got some very bad, we'll isolate them – they'll get double the grub and double the attention.

"The lame ewes are run as a separate flock and foot-bathed at least twice a week and probably turned up at least twice before they go to the ram. And we'll sort those out good and proper. And, as they get better, they'll have a release mark put on them and they'll go back to the flock. And we usually have a hundred and fifty to begin with and we'll cut that down to a hundred. That's seventeen percent [with footrot] – it's quite high. This is a so-called chalk farm – you hear people say, 'they're straight off the chalk!' – but in fact sheep off the chalk *do* have footrot. You're kidding yourself they don't really. You do get footrot but obviously they get more in wet Wales. But you can't be lazy – you've still got to do the job.

"Then, after the foot-paring, we're still running them in their condition groups. The idea is to get them – not too fat and not too lean. Just a couple of weeks before lambing we round-tail [clip the wool off around the tails] all the ewes and that gives us another chance to look at the feet and check the udders.

"This year we fed a lot of hay – we fed about a thousand bales by mid-December. We have to try to gauge the root crops [rape, kale, swedes etc] to last until lambing, and also to suit the farmer who wants to plough it up for his spring wheat. So we have to try and fit in both things: try to suit him because he doesn't want to leave it too late; and trying not to eat it all off so we've got a lot of ewes standing about wondering what we're going to do with them. If we see any very poor ewes, we tend to put them in the barn and give them a bit of cake early.

"And then we do the vaccination, which we have just finished last week. The clostridial vaccination, that is, which is two weeks before lambing. All the new ewes we got in are given anti-abortion vaccines and they are started on the clostridial system. [This is a system of intraperitoneal injections followed by annual boosters against a number of ailments that affect modern sheep.] I think that's fairly common now among flock masters. And they are usually wormed as well, just as a precaution.

"And that takes us to hay and cake, which they are fed, and it's a hard slog: from January right up to April is the feeding, and I have an assistant then from January for six months. He and I undertake to do all the feeding, and flexy-netting of the kale. [This is light flexible netting that is charged by an electric fencer and will restrain sheep. The sheep are moved over the kale etc enclosed in it.] It's quite a lot of moving of fences and organising and feeding. They are only on the kale four hours a day. It makes

it last longer – we ration it."

And then, of course, all Ian has to do is see to the lambing of seventeen hundred ewes, watch their condition carefully afterwards as they graze on summer grass, draught out the lambs as they get fat to sell them, and shear them. I said to him: "Obviously you have a gang in to shear them?"

And he answered: "Oh no – I shear them on contract. I get paid extra for that. And I get time off from the farm to go and shear several other farmers' sheep too. And I have my own to do." For Ian has his own little flock of pedigree sheep which he keeps as a hobby.

To anyone who has done any work with sheep the idea of any man – or two men – "turning up" *seventeen hundred* very heavy ewes – and forty-five even heavier rams – makes the mind boggle.

To an old-fashioned sheep man this whole busi-

ness of high-pressure agribusiness is appalling. To keep so many sheep that you cannot possibly give any of them individual attention, to breed them and feed them so that they have an unnatural number of lambs, to feed both dams and lambs so hard that the lambs mature extremely early; to drive the animals in fact to higher and higher production (Finnish sheep are being introduced which often have quintuplets, and the Friesian, in use now on the farm Ian was working on, is common now) and to fight the inevitable disorders and diseases they get with a more and more complex system of injections and medicines can only be described as being in a pastoral rat-race.

But the land, on which the sheep Ian looks after are kept, is capable of growing wheat. And such is the profit to be made now from wheat forced to very high yields by enormous applications of nitrogen that the keeping of sheep on such land seems to many agribusinessmen an indulgence. All you have to do is: get rid of the sheep – get rid of the shepherd – buy more chemicals – and you will grow more wheat. The land will suffer, yes – but not before our generation has made its fortune – and that is a problem that will have to be faced by future generations.

To see a more relaxed kind of sheep farming –

more small-scale, more organic and more old-fashioned if you like – I went not far away to meet Mr Tom Butcher. Mr Butcher was sixty-seven when I met him; and his employer, whom I did not meet, was over eighty.

Mr Butcher shepherds his sheep at the foot of some steep downland, which is common grazing, and he has the right to turn his flock out on it. They only stay on it so many hours, though, and are then put back in their folds which are either on improved grass or on root crops of various sorts.

Tom Butcher does not use flexinetting but rather the old-fashioned sheep hurdle such as one finds in that part of England. These are made from willow or hazel and are closely woven like a basket. On the Cotswolds, when I was a boy, we kept sheep behind hurdles made of ash, like miniature five-barred gates. These were stronger and lasted longer, but did not have the advantage that the woven hurdles have of being good wind breaks. Tom keeps two or three hundred sheep nearly all the time on arable land, his employer growing a series

of arable crops for them, and his method of shepherding has changed very little down the years.

"I started on the farm 1929 when I was fifteen,
and about twelve months after the Governor
decided to have some sheep. So we went over to
Sixpenny Handley – where they used to have the
sheep market – and he bought about fifty. Me and
another farmer we got them over to the other valley.
We had to walk 'em in those days, of course. And
then next day I took 'em out along the down
without a sheep dog, and the farmer – there's only
him and me you see – he went over to Sixpenny
Handley to get five dozen sheep hurdles. We folded
the sheep in those days because there were no
artificial manures. We relied on sheep manure in
the ground.

"I got on down with my fifty sheep, and there
were eight other shepherds with their flocks. And,
of course, they all had dogs, but I hadn't got one.
And they used to pull my leg, saying if the sheep
had got together I'd lose 'em and all that. But they
were very good – if my sheep were going the wrong
way they'd put their dog around them for me.

"That went on for twelve months and the next
farmer helped me with my lambing the first year
(he had a dog) and after that we got on very well.
I had to turn to and do all the manual work: carry

on the hurdles every day and pitch them out.

"Then the Governor decided we'd have more sheep. We bought fifty each year till we had a flock of about three hundred, and that's as many as I ever looked after: mother sheep – the breeding ewes that is – not counting the lambs.

"All the year round we grow different crops for the sheep to fold on. In the autumn, before we put the tups on 'em in October, we have them on rape and turnips – to flush the ewes before they take the ram. After that they'd go on to a bit of dry grass – any old dry grass, so that you'd steady the sheep up so that they're in lamb and there they stay – they don't have to come round to the ram a second or third time. You've got to flush them up for a month or six weeks first, and then hold them steady on some dry grass after they've had the ram – that's how we've always done it.

"After that we'd fold 'em on grass, all winter round to the lambing. Then we had swedes and kale, folding them all the time. Let them run loose, you see, and they're tramping on everything. If you fold them and give them fresh food every day they've got clean food and they thrive on it. And, if you're fattening sheep, they'll fat better; if they're loose, all the flesh they're getting on them, they're running it off. When they're steady, they'll fatten.

"Every morning we used to take the sheep to Down [turn the sheep out on the common downland]. They're steady those sheep when they get out there; they'd put their heads down and they'd feed their way right round – about a mile and a half – and then they'd start working their way back. You had a dog but you never used it. You never wanted to use it, because the sheep got used to it and they're quite steady.

"At first we had pure Hampshire Downs. They're more of a stocky-legged sheep – not so long

in the leg as a Suffolk or sheep like that. They fat very quick – it's a proper butcher's lamb, a Hamp-shire Down. But, of course, there's not many flocks left now. There's only about half a dozen and they all breed rams and just sell rams from them. And there's no hurdle flock around here now, only this lot.

"Everybody's on this flexinetting, because half of these young chaps can't pitch a hurdle, can they – they don't want to – it's too hard a work. We've got lazy like the rest of 'em now. We're getting older now you know!

"Flexinetting is easier, but you're liable to get the little lambs – they put their heads through. They're not satisfied with what they've got here and they put their heads through and you come up in the morning and perhaps you find a little lamb hung itself or something – dead. I don't like it – it's a lazy way of looking after sheep. Hurdles are harder work, but, now I've got old, the Governor he sends a tractor driver up every day and we just fling the hurdles up on to a tractor and haul them on, so I haven't got to carry them."

As Mr Butcher spoke, I could vividly remember carrying the heavy ash hurdles when I was a boy on Cocklebarrow Farm in the Cotswolds. I could just carry six of them – and this I had been told was

a fair load for a man. I would lean the six up on a leaning stake, push a stake right through them and then heave them up on my shoulder.

I would have to carry them over the muddy pen – the sheep parting before me as I went – climb over two existing fences, and then put them down in the line of the new fence I was erecting for feeding off the sheep the next day. It was very hard work but it never did me the slightest harm. Good work, in the open air, never did anyone any harm: I am quite sure of that.

III

MAN AND DOG:
A RITUAL OF GRACE
AND BEAUTY

TOM BUTCHER mentions his early embarrassment at the lack of a dog. He needed the dog because he used to run his sheep out on the open downland and had to keep them away from other people's flocks – and get them back when he wanted them. If you just keep sheep inside hurdles – or Flexinetting, or in small numbers in small fields, you do not need a dog. In fact an untrained dog can be a damned nuisance.

In South Africa, in the Karoo, we looked after thirty thousand sheep successfully with no dogs at all. But that country is fine for riding over, and our horses could easily out-distance any sheep and the crack of our stock-whips would turn them any way we wanted them to go.

In South West Africa, in tropical Africa, the sheep were shepherded in the ancient arcadian tradition, by native shepherds. The shepherd walked in front of his flock very often and the flock followed him. He needed no dog. You can see this kind of sheep tending in southern Europe to this day. Sometimes the Mediterranean shepherd will have a dog with him, but this is more to keep off wild animals than to herd the sheep. Such a shepherd is always with his sheep and they are not afraid of him.

Only in the British Isles is the working sheep dog brought to its highest state of development. Only in these islands can a sheep master send his dog miles away, out of sight even, to round up hundreds of scattered sheep and carefully bring them to where they are needed. In other countries dogs, if they are used at all, are used just as an adjunct to a man shouting.

Tom Butcher has a dog now, of course, a friendly old hairy thing that does not suffer from over work. Ian Williams has three and is very keen on them: they are symbols of his pride in his ancient profession.

However, to see dogs at their best I went up onto the high fells with Dick Fawcett, his father, an uncle, and eight dogs. We went to round up 600 sheep from 1,100 acres of rough moorland: moor and wilderness. It was summer time – no blizzards – but we started late because in the early morning we could see that, although where we were, in the valley bottom, it was clear and beautiful, up on the fells was mist. Grey clouds rolled over from the south-west.

Eventually we climbed into two Land Rovers and set off through the cobbled streets of Hawes, which is a very gem of a little Dales town, and then up the steep road towards the south-west. We turned off down a rough track, left the Land Rover, and thereafter walked – or scrambled might be a more apt description. Dick's father and uncle had gone along a separate track which would bring them far over to the other side of the moor.

Most of this particular fell consisted of a huge, V-shaped, steep-sided valley, cut no doubt by a glacier in the last ice age. The little tufts of white from the cotton grass showed where the wetter patches were. Heather and ling grew in places; in others dwarf gorse; in yet others coarse grass. An interesting geological feature was the sump holes: very steep-sided holes, often as big as a house, which seemed to have no outlet for water in them. The limestone, being pervious, no doubt lets the water away. But Dick pointed out how difficult it was to find sheep in such places, and how easy it was for a sheep to become buried down there in drifted snow.

There seemed to be very few sheep about. We started one ewe with a big lamb and she ran down the steep hill towards the valley bottom. "We'll gather her in when we come down again," said Dick.

After a couple of miles of rough scrambling, during which it would have been an advantage to have had the right leg longer than the left, we spied some sheep far away on the other side of the valley, high up near the ridge – far far away: we could only just see them.

Dick selected one of his three dogs – Nell I think it was – and called her to sit at his right leg. The other two dogs he told to *stand*. Nell could not see the sheep yet – they were too far away.

"Nell – away! Away! Away!" said Dick

quietly. And away Nell went. Belly close to the ground, as sheep dogs will, she ran. Down the steep slope, finding her way around obstacles, slipping through the bracken, she eventually got down to the valley bottom, crossed the stream down there by jumping from rock to rock, and then we could see her snaking up the steep bank beyond.

Dick now controlled her with a whistle. He had a small flat whistle, about as big as an old penny, which he kept between his lips. The sound it made, although not too loud to us, could be heard by a dog for miles. And then Nell became radio-controlled. Dick could make her do anything he wanted – go left, go right, stand, drive the sheep away, bring them back to us – all by different noises on the whistle. He made Nell drive the sheep out of sight over the ridge, where he knew his father and uncle would be with their dogs. Then he recalled the bitch.

Down we went, eventually, past an old mine working, down down to the valley bottom, crossed the rushing stream by leaping from rock to rock, and started the long steep slog up the other side. At the top of the ridge Richard Fawcett Senior appeared. He is a fit man but by his breathing showed signs of a long slog. His two dogs, too, were panting and their tongues lolled out.

"Have to come up on the Fell occasionally," he told me. "Only way to keep fit." Running the famous sheep auction mart in Hawes means that he does a lot of desk work.

Sheep began to drift along the ridge, the ewes waiting and bleating for their lambs, which were now well-grown – getting close to the size of their mothers. The sheep were Swaledale but the rams with them were Bluefaced Leicester.

We started working our way down the valley again, but this time on the other side of it – and we still needed our right legs longer than our left legs because now we were coming the other way.

Dick stopped and gazed at the other side of the glen through binoculars. He spied two sheep: a ewe and a lamb. It was a long time before I could spot them with the naked eye.

"Zeus! Away! Away!" Zeus sped down the slope. He could not see the sheep at all. His master guided him with the whistle. The sounds he made on the whistle had the same rhythm as the spoken commands: "Away! Away! Come by! Come by! Stand! That will do!"

Again we watched a dog apparently under radio control. The gill was deeper there and when the dog got to the bottom of the valley he had to swim across. The other side was desperately steep and very rough. Painfully Zeus worked his way upwards, heeding his master's whistled commands to bear this way or that way and at the same time having to negotiate obstacles and pick a path through rough terrain covered in barbs by dwarf gorse. He had still not caught one glimpse of his quarry, and was acting solely on his master's word that there was something there.

The stamina of the dog was remarkable. We had already scrambled perhaps six miles over ferocious country and at one point Zeus had been sent far out of sight over a ridge to reconnoitre. The lonely little black and white speck on the other side of the dale – perhaps a half a mile away as the crow flies, but a mile and a half as the dog runs – was visibly slowing down. And still the two tiny white specks – the sheep Zeus was searching for – were oblivious of their approaching enemy.

It is a strange relationship, that of dog and sheep. The dog must look upon the sheep as prey, and yet he knows he must not bite them. The sheep are deadly scared of the dog: the moment any canine comes into view the sheep lift their heads, group together and move away. One of the deepest of all their instincts says: *Danger*! The shepherd can teach the dog nothing new: all he can do is to make use of and particularly control the deep-seated instincts that the dog already has, the instinct to obey the pack leader (in this case the shepherd) and the instinct to herd the multiple prey. We have all seen young sheep dog pups herding ducks, geese, or even children: tucking them in as handlers say, keeping them together, not allowing any to straggle outside the mob.

Sheep are prime practitioners of the flocking instinct – an instinct that baffles animal behaviourists. Obviously, individual safety for defenceless

herbivores should rest in flight and dispersion. A pack of wolves or wild dogs would not know what to do with a herd of deer that scattered in all directions, but deer that keep together in a mob are easily hunted. Konrad Lorenz, who explains so much of animal behaviour, cannot satisfactorily explain this. But if you watch a flock of sheep being savagely dogged you realize that each individual animal appreciates that its safety lies in *not being on the edge of the flock*. The ones in the middle are quite safe. You actually see the sheep on the outside jumping on the backs of others to get towards the middle. It is *always* one of the ones on the edge that gets bitten or, in the case of wild animals, killed. And thus so many humans seek the anonymity of the crowd.

I remember when I was a learner in the Karoo in South Africa being sent off on a horse to bring back a flock of a few hundred young merino rams. I cantered round them, rounded them up (we did not use dogs on that farm) and began to bring them back. But one young ram, and one only, showed individuality. He would *not* stay with the flock – he broke out of it again and again. Finally my horse and I both made the same decision: to discipline him.

Horses which are used to herding animals derive great fun from it and are extremely skilful at it. Instead of driving the errant ram back into the flock, as we had been doing, we kept him from getting back – we drove him away from the flock. Now the ram's attitude changed completely. He wished desperately for that anonymity again. He wanted back, but we wouldn't let him. Turn and twist as he would, my horse was quicker and – only twenty as I was and not very humane – I was always ready with a cut from the *sjambok*. The horse and I kept this up too long. It became a game and we overdid it. The ram suddenly dropped to the ground – apparently dead.

Now these rams were very valuable. My boss bred stud rams for sale; they were pedigree and expensive. I had been warned many times not to press them too hard in the hot African sun: their hearts could go. I got off the horse and tried to get the ram up on his legs. He was inert – seemingly quite dead. I lifted his head and it just flopped back again. I stood for a few minutes wishing I had not been so foolhardy. Then I had an idea, mounted my horse and cantered after the rest of the flock. I drove them back over the body of the ram. Then I slowly drove the flock away in the direction of the farm again. And behold, there was no dead ram! The disciplined animal had gained the anonymity of the flock again, and I will wager he never again made any bids for independence. Thus do totalit-arian regimes discipline their citizens. Orwell's *1984* is about just that.

But to get back to the Yorkshire hillside – a far cry from the level, shrub-covered plains of the Karoo – the dog Zeus was guided by Dick's whistle to within a few yards of the two sheep before he saw them. Then no further control was needed. Zeus took over: he knew exactly what to do, which was to bring the quarry back to his pack leader.

By now sheep were pouring over the top of the ridge and coming down towards us. They would no doubt have broken back up the glen if Zeus, Nell and Bob had not been there to stop them.

Close by there were some ancient enclosures, now in ruin – walls of dry limestone, the crumbling foundations of a house, a stone-walled pen – and the land, fairly level just there, looked as if it had once been improved. The grass was better than on the moorland.

"This used to be where the sheep were lambed," said Dick. "A man would live in that house for the lambing. We take them down to the dale now."

We were still high up in the mountains, and could see the flat floor of Wensleydale's misty-blue far away below us. Dick's uncle appeared over the brow of the hill with his two dogs. Now with five dogs, we began gently to move the ewes and lambs downwards again. "Happen we haven't missed many," said someone.

Much further down we came to a fence: the whole eleven hundred acres of the fell were enclosed by it. We let the sheep pass slowly through a gate. There was a great bellowing of ewes calling for their lambs and vice-versa. The dogs were carefully kept back. "Stand! Steady! That'll do Bob." The sheep were not to be hustled through the gate.

Dick Fawcett is a champion trials dog handler and wins many prizes. He likes talking about dogs too.

"I have got three mature, fully-trained dogs, which both trial and work, and one seven-month bitch which is about three-quarters trained, and one four months which is just starting to play with sheep, and a litter of pups. This is what I try to do – is to have half a dozen, and I find if I have any more I don't do them justice, and if I have less there is a danger – I like to have young ones coming along each year you see. There is a nursery season each winter for young dogs, so I like to have a young one coming along each winter. There's a danger – if you don't have enough coming along and one turns out to be no good, you've had it.

"I like to have three mature ones and three young ones. This is for trials work – the three dogs I have can cope easily with the work I have, but they have got to be able to do both – or they have got to go. And the two don't necessarily go together.

"I go trialling most Saturdays of the year. And, when it gets round to August, sometimes I feel a bit guilty. Mind you, I might fly off about five in the

morning – run my dogs – then come straight home
and put my full day's work in after that you see. You
don't have to be there at a particular time – if you
want to get there early, you can.

"The nearest trial ground we have here is six

miles away. Next Saturday I am going the other
side of York which is seventy miles. If I'm milking,
I try to milk as early as I can and then go.

"But the dogs are running particularly well this
year, so I try to run them as often as I can. It's like
this: if things are going particularly well, you are
more keen to go. And I am very very ambitious –
one day I would like to be, dare I say it, one of the

top men. But that's something that takes a long time.

"You don't pay much for a good working pup – about thirty to fifty pounds. The very best working dogs are up in the thousands now – that's the very best – but a good working dog: three – four – five hundred.

"When they get to about eight or nine years old they start going downhill after that. A lot depends on the dogs. I have got three dogs now and two of them never stop running – they run a lot of the time when they needn't be running. But the best dog has got more up here and he *thinks* about it; and I

should think he will last a lot longer than the other two because he is not thrashing himself to bits all the time.

"There are no party tricks involved – they are all practical commands, like they have a command for left and a command for right. I have a whistle and voice command and I have different whistle and voice commands for different dogs, so I can run two at the same time. On the fell – if they are all on the same command – you can't have them running together because there are a lot of things – because one dog might be doing something and another something else. You have a command for right and left – fetch to me! – drive them away! – stop! – shed! (that means take a few out of the bunch). We couldn't work sheep on the fell without the dogs.

"It all depends on the dog. It's a natural instinct – you could never train a dog until it wants to work with sheep. If it's not there, you can't put it there. Some dogs don't work until they're about two years old. The one I have now which is seven months old – when I got it to eight weeks, which is the correct weaning age, it was sufficiently interested in sheep then to approach them in a mature fashion – not running around them with its tail up in the air barking, but walking up to them and popping round them and that sort of thing. I mean a lot of farmers would think that was fully trained. Now to me it's nowhere near it – it's got to be polished now. They are like people. Some are thick and some are bright. Some just do it naturally. I try to avoid walloping them but they all probably get walloped once in their lives. They are like people: some will take it and some won't, and some you only need to growl at them and they will behave themselves. Some are thick enough and they are no good to me, so I get them to a saleable type and then sell them. They are always good to sell – you could sell a fully broken dog every week."

And how do you train a dog to answer to some half dozen commands; to distinguish these from the commands you are at the same time giving to another dog; to obey your commands three-quarters of a mile away and even when out of sight; to retain his own independence and individuality; to use his inborn hunting instincts; and to really hunt down his prey, yet never overstep the mark?

"Well, when I take them out walking we go along, just talking to them, and, I don't know, just slap my leg and say: 'That will do!' and they just come to you; and then they start associating 'That will do!' with 'Come to me!' and then I start to use the whistle – you need a whistle over a long distance. I can't whistle very well – I use a tin whistle: not many shepherds use their fingers any more. I can make any command I want on that.

"A funny thing happened to me the other day: I sold a dog to a Welshman, and he took it home with him and the dog just wouldn't work for him. So he wrote and told me. So I sent a tape down of my commands, and he started to run like a hero but the moment he turned it off the dog just stopped. He kept her for a month and then she came home – and she worked like a hero! I kept her for so long and then a friend took her, to see if he could get her going, and she would work until he gave a command and then she would just stop – she would only work when she wanted to. And then this lady came along and she said she was just buying a farm and that she needed a dog – and she was a real lady's dog; a nice kind of a nature she had. The lady fell in love with her and she wanted to buy her, so I told her what had happened, and I said, 'If you buy her, you buy her on the understanding that she doesn't come back again, because it's not going to do me or the dog any good if the thing goes to and fro all of its days.'

"She thought about it and decided to buy her. But several months later I got a telephone call and she seemed in a terrible state. She said the dog wouldn't work and had already killed two chicks and one cat! But eventually she got her working; and I spoke to a friend over there who said she was going like a good 'un now! She was delighted with her now and it was just a matter of getting used to someone else.

"The dogs are my second interest to my wife Anne. When I am working I am always thinking of ways I could improve them or negotiate the next trial: I have usually been there before so I know the place.

"They will work a long way from you and even out of sight; but, if it's possible, it's better if they can see you because they might be gathering a lot of ground you don't want them to gather. They wouldn't do any harm because they are not that sort of dog, but they might be exhausting themselves doing things you don't really want them to do. So I suppose I can work half to three-quarters of a mile away from me, so long as it isn't a terribly windy day and they can't hear you.

"You forget the difference between a dog and a human. You might think – oh there are some sheep over there and it is quite possible the dog can't even see the sheep because they are down there: but, if you get to the dog's level, you realise the dog can't see anything; and you wonder why the dog didn't go in the right direction!

"Our fell is difficult. It has big gills. You could send two or three dogs up and eventually they would meet, but one dog on its own wouldn't be able to hold them. Anyway – the dogs need guiding. It is like driving a steering wheel. They won't necessarily take them down the easiest way – they might take them anyway – so it would be impossible to

send them alone. The top of the fell is a lot of limestone – great big rock steps, big flat tables – and then
on the highest of them all there is millstone grit and
peaty wet ground. As you come down the gill, it is
pretty sweet grazing down there. But there is a lot
of peat – it could stand a lot of drought that peat
could and still be wet. There is very little heather – it
is all bilberries and cotton grass. The rainfall up
there would be eighty inches."

Dick's description of his fell land allows one to
envisage just how rugged and severe the ground is
over which his dogs have to work.

It is very difficult to get sheep-dog handlers to
explain to you how they train a pup. It is something
instinctive, that either a man can do or he cannot
do. It has something to do with intuition, I believe,

and extreme sympathy and community of interests between the two hunting animals: man and dog. But Dick has a good try at explaining:

"I take my pups out as soon as they can walk and I just get them used to not standing on my feet. It is all learning. I just give them quite a kick out of the way – not a kick, but just enough to let them know they haven't got to be running in front of you, which is normal for them. And then to be able to negotiate steps and stiles and things – it's all a great advantage to have all of this behind you for when you start training them with sheep.

"And just chatting with them, and teaching them their name. And I suppose any time after that I teach them to stop. I tell them to, 'Stand!'. It can be 'Sit!', 'Stop!', 'Stand!', so long as they stay where you want them to stay. I always use the same word for all the dogs, and I say: 'Stand!'

"That there again takes time, when I have had my supper and finished my work, and I feel like going out for half an hour or so. And then I just take them to some hoggs – I always have some hoggs there just handy, in a small paddock. Then I take the pups there one at a time, and sooner or later they will prick up their ears and set their eyes and you are away you know! It has started. Then it is up to you to encourage it and try very carefully to train it. It has got to be done very carefully, because otherwise you could dishearten it. If you start making them do something and they are not sufficiently keen so you stop them doing it, they then think: 'Oh he doesn't want me to look at those sheep!' and he goes and cowers in a corner or something.

"You have got to determine whether it is a strong-willed dog or a little sensitive thing, and then you go from there. The one I have now which is four months old – he likes to take them to the wall and hold them there, but he won't go round them. But he's only been doing this for about a month now. And there is no way that I am going to be able to make him go round them now in these little paddocks, so I am going to have to wait until we've finished the hay and got the grass out of the bigger fields. Then I'm going to put a bigger flock in there which will stand in the middle – and the sheep aren't going to move if they are in a big block – and it will learn to go round them then.

"Well nearly all dogs are lop-sided, like we are right-handed and left-handed, and they prefer to go round sheep on that one side of them. If he is left-sided then I either say: 'Away!' or 'Keep off!' – I have two different commands. And every time he would go round them I would keep saying: 'Away! Away! Away!', all the time he was doing it, and then call him back and do the same again, and say: 'Away! Away! Away!'. And in time he would realise that he was meant to go left.

"And then you have got the hard part really, and if it's lop-sided then you have got to get it to go *right*. (They're not all lop-sided mind you.) You shouldn't be doing it unless they are sufficiently wanting to be trained and if he *does* want to be trained then he will allow you to make him go right. Well, you get the dog to stand on the right-hand side, and if you let him go he will cut in here and go round to the left, if he is left-sided; but you've got to make sure he doesn't so you have to run forward and, once he gets near to the sheep, he has no choice but to go right. And the thing is – if it is very lop-

sided – to concentrate on the other side until the two sides are equal. I say: 'Come By!' or 'Go Ahead!' for that command. I mean everybody has different commands.

"Some dogs will do it better than others. Now, if Zeus gets behind sheep then – it can be any distance away – he will bring them to me. Nell gets behind sheep, she hasn't what we call a sense of balance. She will take them where they want to go. She will stay behind them though. So, if they are pointing in the wrong direction, she will follow them. Whereas Zeus will balance them: he will tuck the sides in just as you tuck your bed in, get them together, get behind them and bring them straight back. But Nell will let them go all over the place.

"Now, they will all get behind them in time and they will bring them to you, but the most difficult thing about teaching a dog and that's driving them *away*, because that is a foreign thing to them: that's the herding instinct back-to-front. That often takes the most time of anything. So what you do is: you get the sheep going and you walk them behind the sheep and then you make sure the dog stays there instead of popping round them. And every time he starts popping I say: 'There!' – that means, 'you have to walk up to them *there* in the position you are now!' And you gradually get them used to it, until maybe you can stand still and *they* will walk a few strides of their own. And slowly they will learn to do it by themselves.

"This is the crunch really. I suppose you need a lot of patience really and I don't think I am a particularly patient person, but I do seem to be with dogs. Because I get some pleasure out of it – I get a tremendous amount of satisfaction out of them when I see them progressing!

"I don't fall out with them, but I reprimand them when they are being naughty, and I pull their ears if they are not doing as they are told. I get their ear like this and I pull it. It's usually a matter of when they haven't stopped when they are supposed to, which I suppose is the biggest fault; and I shout in their ear: '*Stand*! *Stand*! *Stand*!', and then gradu-ally I let the ear go down and I carry on saying 'Stand!', and then I start to stroke him and say 'Stand!' until we are big pals you see. He has realised he has got to stand, but once I finish the punishment we are still friends, so the next time he goes off he knows when he has got to stand and he is not being naughty because there is still this bond between us. That's just my system and it seems to work, but a lot of people seem to fall out with young dogs and spoil them by hitting them too much and by expecting too much too soon.

"I never call a dog to me to punish it. I don't think that is good policy. I go to it, and if it starts running away then there is no point in doing it anyway, because it has realised that it has done wrong so there is no point in punishing it. Just the fact it knows it has done wrong – if it is going to be any good, it won't do it again you see. There is no good in punishing for punishing's sake – you must be doing it for a reason."

To watch Dick Fawcett training his pups and young dogs in the small paddock near his house, where four young sheep are kept for that very purpose, is to understand partly some of the above remarks. The dog snakes towards the sheep, belly close to the ground, in that ominous way that sheep dogs have – hypnotising the sheep with his eyes – but at the same time listening to the commands of his master, casting a look in his direction from time to time out of the corner of his eye to see if he is gaining approval.

The dog wants desperately to please his master. If the man gets pleasure from controlling the dog, the latter gets an even intenser pleasure from responding correctly to these controls. The two mammals, man and dog – two predators really – work as one machine. The dog knows that the successful pack animal must obey the leader of the pack. The man knows that he needs the dog desperately, that he cannot earn his living without him. It is a hopeless task to try to round up sheep from a rough mountain area without well-trained dogs. I have tried it. In Wales I have set out to gather sheep with neighbours who owned badly-trained dogs and the enterprise was a complete failure. Men were running everywhere, shouting and waving their arms, and the sheep were breaking back and just laughing at us. One man with one well-trained dog would have done the job with no shouting at all.

Sheep in many other parts of the world are never out of the control of their owners. They are penned up at night while the shepherds sleep, and during the day they are in the close control of the shepherd all the time. Such sheep are truly domestic animals.

But the sheep that range over the high hills and mountains of the British Isles are feral animals. If they see a man, they run away from him. For months sometimes they never see a man, or a dog. They have to be *hunted*. A hundred men on foot with no dogs would not have a hope of gathering in the sheep on the Fawcetts' fell. A dog could not do it alone – nor a pack of dogs. But the man with one good dog can do it easily. And to watch it

being done is like watching a superb ballet, a work of high art, a ritual of grace and beauty.

Let me give young Richard Fawcett of Wensley-dale, who has helped me so much in this attempt to explain these esoteric arts to laymen, the last word:

"I like sheep and I have a lot of pride in trying to produce the best. I like to be best at whatever I am doing – I don't suppose I ever will be but I still like to try to do it. And I have a lot of interest in trying to find a tup that's going to suit our sheep and produce the lamb we're looking for.

"I prefer working with sheep to working with

cows, because cows are there twice a day to be milked and to be mucked out, and it becomes a drag, I think, does milking cows.

"Sheep work is seasonal – there are times when you don't do much to them. And I don't think of it as work when I am working with sheep really: if I am going shepherding, it is *shepherding* – you know – it is different, isn't it?

"Well that is how I feel about it – I *love* going up there!

"*But you see: even if you only have one missing you have got to go there and have a look – well if you are worth anything – the same as if you had a hundred missing.*"